Y0-EMP-306

WITHDRAWN

Field Trips

At the Zoo

By Nick Rebman

level
2
little blue
readers

www.littlebluehousebooks.com

Copyright © 2020 by Little Blue House, Mendota Heights, MN 55120. All rights reserved. No part of this book may be reproduced or utilized in any form or by any means without written permission from the publisher.

Little Blue House is distributed by North Star Editions:
sales@northstareditions.com | 888-417-0195

Produced for Little Blue House by Red Line Editorial.

Photographs ©: FamVeld/iStockphoto, cover; Trong Nguyen/Shutterstock Images, 4; luckyraccoon/Shutterstock Images, 7 (top), 24 (top left); ChameleonsEye/Shutterstock Images, 7 (bottom); Romrodphoto/Shutterstock Images, 9 (top); Tharin kaewkanya/Shutterstock Images, 9 (bottom); Ekaterina Malskaya/Shutterstock Images, 11; Aris Suwanmalee/Shutterstock Images, 12 (top); Super Prin/Shutterstock Images, 12 (bottom); Sukpaiboonwat/Shutterstock Images, 14–15; Carlos Sala Fotografia/Shutterstock Images, 17, 24 (bottom left); Q-lieb-in/Shutterstock Images, 18, 24 (top right); Charlie Unggay/Shutterstock Images, 21, 24 (bottom right); Khoe/Shutterstock Images, 23

Library of Congress Control Number: 2019908246

ISBN
978-1-64619-033-1 (hardcover)
978-1-64619-072-0 (paperback)
978-1-64619-111-6 (ebook pdf)
978-1-64619-150-5 (hosted ebook)

Printed in the United States of America
Mankato, MN
012020

About the Author

Nick Rebman enjoys reading, walking his dog, and traveling to places where he doesn't speak the language. He lives in Minnesota.

Table of Contents

At the Zoo

We go to the zoo.

We want to see many

different animals.

We see a monkey.

It eats fruit and plays on a rope.

fruit

We see giraffes.

We give them food.

We see goats.

We give them food.

giraffe

goat

We see a snake.

We can hold it.

The snake is long and thin.

snake

Many Colors

We see many colorful birds. Some birds have pink feathers, and some birds have blue feathers.

We see zebras.

They have black and

white stripes.

They eat dry grass.

We see turtles with

green shells.

The turtles rest on logs.

Big Animals

We see an elephant with long tusks.
The elephant sprays water with its trunk.

We see a tiger.

It has sharp teeth and

long whiskers.

whiskers

We see three hippos.

They are large and heavy.

They eat green plants.

Glossary

fruit

tusk

shell

whiskers

Index

31901067115552